A BUILDING CONSERVATION GROUP PUBLICATION

CW01305561

A Checklist for the Structural Survey of

Period Timber Framed Buildings

Text and Illustrations by
David J Swindells and Malcolm Hutchings

RICS BOOKS

Prepared for and on behalf of the Building Conservation Group
of the Royal Institution of Chartered Surveyors

Published by Surveyors Holdings Limited
a wholly owned subsidiary of
The Royal Institution of Chartered Surveyors
under the RICS Books imprint
12 Great George Street
London SW1P 3AD

Reprinted with amendments January 1994

ISBN 0 85406 543 1

Designed by Alan Bull
Typeset by Columns Design and Production Services Ltd, Reading
Printed by Hawthornes Printers, Nottingham

Contents

Foreword

It is intended that this booklet should be used as a checklist or aide-mémoire when carrying out a structural survey of a traditional, historic, timber-framed building of not more than three stories, of the type most typically associated with the pre-Georgian period. It suggests a logical method of carrying out an inspection of a structure so that hopefully no points of importance or significance are missed in the survey report.

The person using this booklet is assumed to be a chartered surveyor, or other suitably qualified person, with some previous experience and knowledge of timber-framed buildings. Although intended and designed mainly with the inspection for potential purchasers of domestic buildings in mind, it could be useful on other timber-framed structures such as barns.

Before actually approaching the building with this booklet you should have gone through all the normal initial stages of taking instructions, confirming instructions with the client, making an appointment with the vendor and agreeing a fee, etc. You should also arrive at the site with the usual tool kit, which will include a sharp pointed implement of some kind. We do not intend to go into further depth with regard to tools or pre-inspection preliminaries as they are beyond the scope of this booklet. The *Books for Further Reference* section suggests other sources where this information can be obtained in greater detail.

Investigations into the geology of the area should be made before going to the property and, on arrival, familiarise yourself with the local topography. Have a walk around and note the type and position of trees and drains etc. in relation to the building. Once you have done this the suggested inspection routine follows the method we use when carrying out an inspection – namely a general walk round to get the "feel" of the property and then starting on the survey itself, elevation by elevation.

You will note that the booklet has been printed on water-resistant paper so if it gets wet on site it can be hung up to dry. It also has a stiff cover and wire binding and has been printed on only one side of the page for ease of use in windy conditions, as surveys of property seldom seem to take place in good weather!

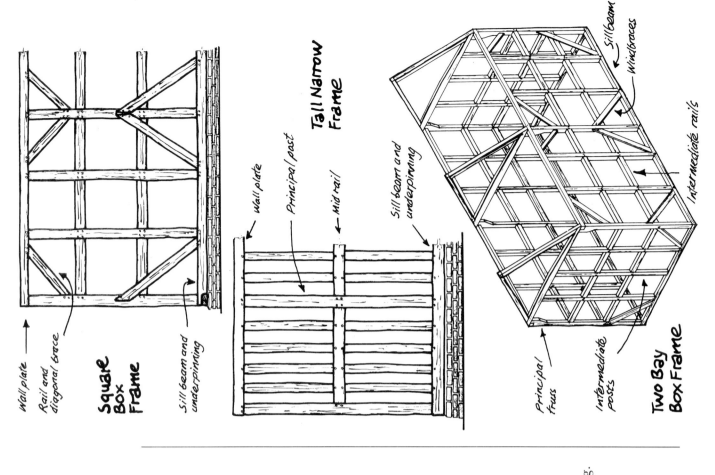

Square Box Frame

Wall plate

Rail and diagonal brace

Sill beam and underpinning

Tall Narrow Frame

Wall plate

Principal post

Mid rail

Sill beam and underpinning

Two Bay Box Frame

Sill beam

Windbraces

Intermediate rails

Principal Truss

Intermediate posts

Walk Round

Check the type of frame

Is it
1. oak?
2. black poplar?
3. elm?

Is it
1. a box frame?
2. fully framed?
3. part framed?
4. a cruck construction?

Are the frames and panels
1. square?
2. tall and narrow?

How many bays are there?

Has the building been
1. extended?
2. constructed with re-used timber?
3. part demolished?
4. previously repaired or altered?

Look for old openings and check
1. windows and doors.
2. new openings and modifications.

Pay attention to the
1. sill beams and their relationship to principal trusses.
2. south and south west elevations for signs of weathering.
3. north elevation and look for any constructional differences.

1

The Survey

External walls

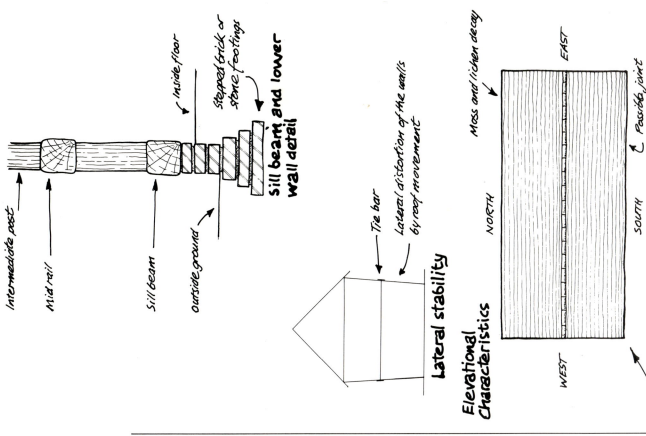

Intermediate post

Mid rail

Sill beam

outside ground

inside floor

stepped brick or stone footings

sill beam and lower wall detail

Tie bar

Lateral distortion of the walls by roof movement

lateral stability

Elevational Characteristics

Moss and lichen decay

NORTH

EAST

WEST

SOUTH

Possible joint shrinkage and decay

Direction of predominant weather

Check the

1. position and design of the sill beam.
2. relationship of the outside ground levels to the lower frame.
3. condition of internal wall panels and timber near the sill beam.
4. principal post junctions with the sill beam for decay.
5. mortice and tenon joints and the trenail pegs.
6. construction to see if the principal trusses rest on the sill beam.

Also look for

1. signs of replacement scarfing to the sill.
2. signs of structural failure and the effect on other parts of the building.
3. evidence that the sill beam has been replaced by other material, e.g. concrete, brick.
4. decay in the intermediate posts and struts.
5. decay in the trenail pegs and repair with mortar or metal.
6. consecutive carpenters' marks.

Stand back and look at the frame check

1. lateral stability and evidence of principal truss decay causing vertical instability.
2. for signs of tie bar restraint and other metal repair to joints.

On the south side check

1. joint shrinkage.
2. loose or missing trenail pegs.
3. internal water ingress at panel junctions.
4. the condition of rainwater goods and proximity to the wall plate.

Tie bar restraint and metal repairs to joints

Angle bar

Round bar and fixings

Metal plate repairs

Mortice and tenon joints

Typical post and rail junction

Typical post and sill beam junction

Frame distortion and spread

Spread

Single metal grap plate

Typical tie bar plates

External walls

On the **north side**

   look for

1. signs of moss and lichen decay.
2. internal frame damage by long-term damp.

   Also

1. check internal panel condition for decay at the frame junctions.
2. look for open shakes in the wood grain and evidence of repair.

   Generally ask

1. have the timbers received preservative treatment?
2. are there any current guarantees?

If there are guarantees

get these validated prior to exchange of contracts.

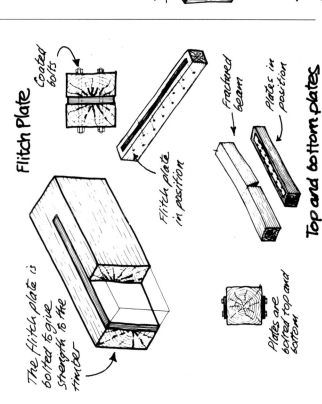

Flitch Plate

Coated bolts

Flitch plate in position

The flitch plate is bolted to give strength to the timber

Fractured beam

Plates in position

Top and bottom plates

Plates are bolted top and bottom

Clap board

Mathematical tiles

Brick cladding

Tile clad

Pargetting

Stone cladding

Typical external wall coverings

Wall panel

Vertical post

Sill beam over- an
inserted DPC

Brick underpinning

Sill beam and dampcourse detail

Tie bar
plate

Face of
external frame

Tie bar

Joist

Trimmer

Means of lateral support

External wall panels

Are the wall panels

1. wattle and daub?
2. brick and if so in what bond?
3. stone; what type of stone and how suitable is it?
4. fully clad?

WATTLE AND DAUB

Check the

1. junction with the frame for evidence of water seepage.
2. extent of hair cracking.
3. finish (if any); is it lime washed or painted?
4. condition of flashings.
5. position of window sills and rainwater drips for penetration dampness.
6. position of any damp course beneath the sill beam and its relationship to cladding panels.
7. position of ground levels against the infill panels.
8. outer surface for signs of wattle decay.

BRICK PANELS

Check the

1. junction with structural and intermediate frames.
2. mortar bonding; are the bricks bedded in lime or sand/cement mortar?
3. stability at infill panels and look for lateral movement.

Is the frame

1. vertically distorted by the weight of the infill panels?
2. laterally distorted at roof levels?

Look for

1. timber replacement/reinforcement.
2. metal tie bar restraint.

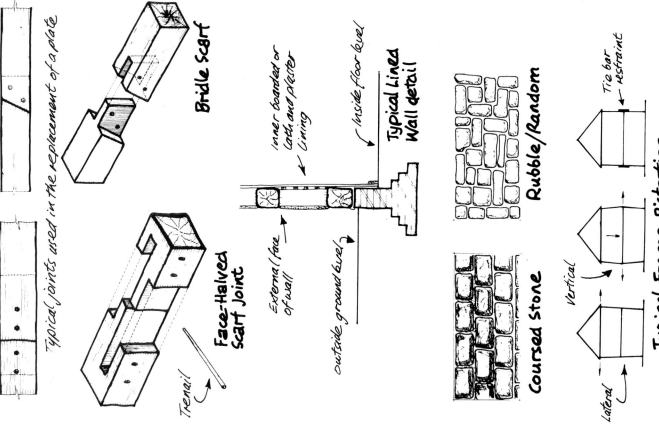

Typical joints used in the replacement of a plate

Bridle Scarf

Face-Halved Scarf Joint

Trenail

External face of wall

outside ground level

Inner boarded or lath and plaster Lining

Inside floor level

Typical Lines Wall detail

Rubble/Random

Coursed Stone

Typical Frame Distortion

Tie bar restraint

vertical

lateral

If in doubt	recommend back up advice from a Chartered Structural Engineer.
Check the	1. wall thickness at each floor level.
	2. provision of lath and plaster inner linings or boarding.
If the **walls are lined** *look for*	1. inner plaster and lath decay.
	2. decay to boarded linings at each floor level.
Look for	1. signs of metal tie failure.
	2. brick and mortar spalling.
Comment on	the need to repoint, repair or reconstruct.
At the **south elevation** *look for*	1. shrinkage of joints near principal posts and rails.
	2. signs of penetration dampness and mortar erosion particularly at the south west corner.

STONE PANELS

Are the stone panels	1. coursed stone bedded in lime mortar/sand cement mortar?
	2. rubble?
Are the internal frames	1. exposed?
	2. covered with lath and plaster?
	3. covered with an internal skin?
Is the frame	1. vertically distorted by the weight of the infill panels?
	2. laterally distorted at roof level?

slate/stone insert

Vertical struts

Sill beam or rail

TYPical Northern Slate/Stone panel

Timber boarding

Mathematical tiles

Brick

Vertical tiles

Render/Pargetting

Stone

Fully clad Frame coverings

Look for
1. timber replacement/reinforcement.
2. metal tie bar restraint.

If in doubt
recommend back up advice from a Chartered Structural Engineer.

Check
1. stone erosion and spalling.
2. mortar erosion.
3. the relativity of ground levels to lower panels.
4. the provision of lead flashings and damp proof courses (DPCs).

In northern areas *look for*
1. tall narrow panels with segments of slate or other local stone wedged between.
2. movement in the frame due to weight distortion.
3. water ingress and decay at the structural junctions.

FULLY CLAD FRAMES

Check the type of cladding

Is it
1. vertical tile hanging?
2. clap or weather boarding?
3. render/pargetting?
4. mathematical tiles?
5. full stone cladding?
6. brick cladding?

11

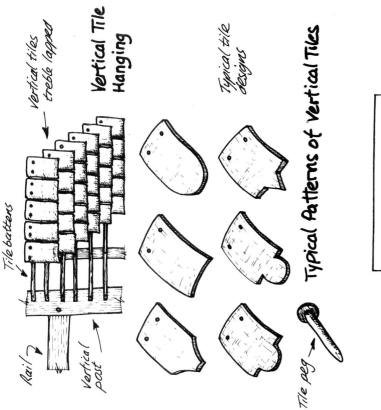

Vertical Tile Hanging

Typical tile designs

External wall panels

Tile battens

Rail

Vertical post

vertical tiles treble lapped

Tile peg

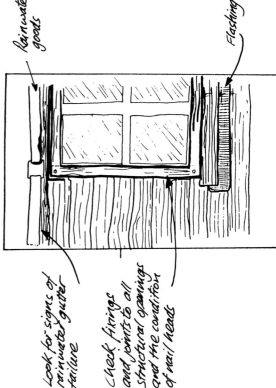

Typical Patterns of Vertical Tiles

Rainwater goods

Flashing

Look for signs of rainwater gutter failure

Check fixings and joints to all structural openings and the condition of nail heads

Timber / Clap Board Cladding

Vertical tiles

Check

1. whether the tiles are hand made or machine made.
2. the line of tiling relative to the roof eaves; this may show evidence of vertical distress in the structural frame.
3. the door and window openings, evidence of weathering to edge flaunchings and the provision of flashings.
4. the frequency of tile nailing.
5. for any tile slippage.
6. hogging movement in the frame and tile covering.
7. the positioning and fixing of rainwater downpipes and gutters.
8. the rainwater downpipe outlet at ground level; is it connected to a shoe or soakaway?
9. the extent of moss and lichen on tiles.
10. the provision of flashings under window frames and their condition.

Clap board

This is commonly located

1. in the south east of England.
2. on residential and agricultural buildings.

Identify the material used

Is it

1. oak?
2. elm?
3. creosoted/painted softwood?

Lath and pargetting cover-to frame. Wire mesh can take the place of laths

Vertical studs

Plaster and Pargetting

Outside face of render

The drip is concealed behind the render face. Moisture will penetrate here

Outside face of render

Poorly formed drip

Rainwater Drip Details

External wall panels

Check
1. for signs of hogging in the boards relative to the structural frame.
2. the type and age of fixing nails; old nails may not be galvanized.
3. the junction of the boarding with window and door openings and look for flashings and condition.
4. eaves boards for decay from faulty rainwater goods.
5. rainwater gutters and downpipes for board decay due to leakages.

*If **decay** is noted*
recommend opening to see the structural frame, subject to gaining local authority/statutory consents where necessary.

And look for
1. plants and shrubs at the base wall.
2. high ground levels.
3. signs of timber preservation treatment.

Render and pargetting

Look for
evidence of sub-structure.

Is it
1. brick?
2. stone?
3. wire or mesh?
4. wattle?
5. lath construction?

*Has the **wall***
1. dropped vertically?
2. moved out laterally?

Face tile

corner tile

Typical Mathematical Tile shapes

Bedding Joint

Face of tiles

Bedding Joint

Batters

Typical section of tiles

Mathematical tile details

External wall panels

Due to	1. failure of structural posts?
	2. movement in the roof?
	3. water ingress from behind the panels?
Due to	1. cracking?
	2. spalling?
	3. failed rainwater goods?
	4. live render?
If so	check inside for dampness and timber decay/insect attack.
Also check	1. junctions with doors and windows and all flashings and fillets.
	2. the adequacy of window sills and door thresholds to the outside face of rendering.
Check	1. that window sills project beyond the face of render and that a drip is provided.
	2. that door sills properly project and are above the outside ground level.
If not	1. warn of potential rising and penetrating dampness.
	2. check all internal frames for decay nearby.
Check	1. decorative surfaces for condition.
	2. for signs of historic failure and cover up repair.
Mathematical tiling	
Note	1. the pattern of tiling, particularly detailing at structural junctions.
	2. whether they are machine made or hand made.

Wall panel can be tile or brick

cover board

The cover board can be used as a flashing

Cover board

Frame

Frame

Inner lining

External Cover Board Details

Chipped and broken/slipped slates can cause batten decay and failure of the rafter feet

Sign of nail sickness

Slate Hanging

External wall panels

Be sure to check	1. the south west and south elevations for extra wear and decay.
	2. any visible framing nearby.
	3. all flashings and fillets.
	4. the fixing detail at eaves level.
Look for	1. signs of secondary fixing.
	2. evidence of sand/cement mortar repair.
If this is noted	1. the battens may have failed.
	2. the frame or essential joints may be suspect.
Check	1. rainwater gutters and downpipes for proper fixing and signs of leakage.
	2. ground levels.

Slate hanging

This is commonly located	in slate bearing areas.
Look for	1. nail sickness.
	2. slate dropping near the eaves.
	3. evidence of hogging.
	4. batten decay.
	5. condition of all flashings.
Check	1. joints at quoins; particularly the south west quoin.
	2. batten ends where visible.
	3. all fillets and flashings to window and door openings.
	4. rainwater grooves to the underside of door and window sills.
	5. internal walls with a damp meter particularly where they are dry lined.

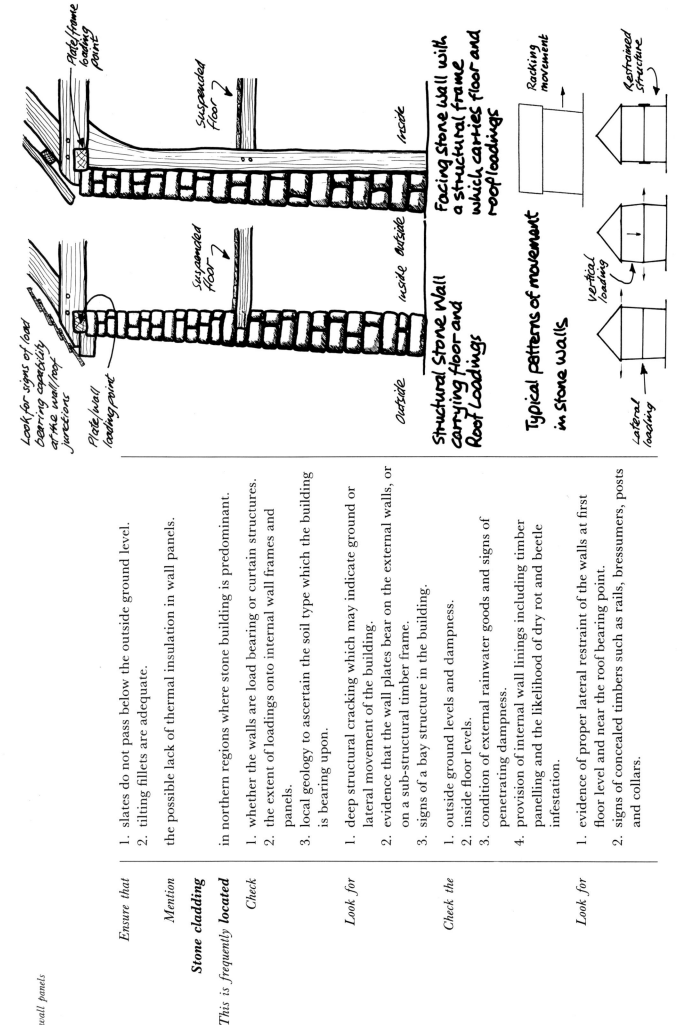

Diagram labels:
- Plate/frame loading point
- Suspended floor
- Inside
- Facing stone wall with a structural frame which carries floor and roof loadings
- Look for signs of load bearing capacity at the wall/roof junctions
- Plate/wall loading point
- Suspended floor
- inside / outside
- Outside
- Structural Stone Wall carrying floor and Roof Loadings
- Typical patterns of movement in stone walls
- Racking movement
- Restrained structure
- vertical loading
- Lateral loading

External wall panels

Ensure that

1. slates do not pass below the outside ground level.
2. tilting fillets are adequate.

Mention

the possible lack of thermal insulation in wall panels.

Stone cladding

*This is frequently **located*** in northern regions where stone building is predominant.

Check

1. whether the walls are load bearing or curtain structures.
2. the extent of loadings onto internal wall frames and panels.
3. local geology to ascertain the soil type which the building is bearing upon.

Look for

1. deep structural cracking which may indicate ground or lateral movement of the building.
2. evidence that the wall plates bear on the external walls, or on a sub-structural timber frame.
3. signs of a bay structure in the building.

Check the

1. outside ground levels and dampness.
2. inside floor levels.
3. condition of external rainwater goods and signs of penetrating dampness.
4. provision of internal wall linings including timber panelling and the likelihood of dry rot and beetle infestation.

Look for

1. evidence of proper lateral restraint of the walls at first floor level and near the roof bearing point.
2. signs of concealed timbers such as rails, bressumers, posts and collars.

Roof construction

Crown Post

Upper collar · *Crown braces* · *Rafters* · *Wall plate* · *Principal Post with tapered Jowl* · *Sill beam* · *Windbrace*

Arched Braced Collar Beam

Ridge piece · *Principal rafter* · *Trenched purlin* · *Arch brace* · *Wall plate* · *Principal post and Jowl*

Check the design of the roof

Is it

1. Crown Post?
2. Arched Brace Collar Beam?
3. Trenched Purlin?
4. Cruck?
5. King Post?
6. Single Collar Rafter?
7. Clasped Purlin?
8. Queen Post or any other classification?

Look for

1. evidence to indicate the likely age of the building.
2. signs that the roof may be a later construction.
3. evidence of structural failure due to decay.
4. evidence of lateral movement in the structure.
5. solidity in the main structural joints.
6. smoke blackening.
7. consecutive carpenters' marks.

Check

1. as many accessible timbers as possible with a sharp implement, having respect for the historic fabric.
2. for movement and the need for further support in the purlins and principal collars.
3. rafter feet at the bearing point above the plate.
4. for signs of insect/fungal decay.
5. for evidence of live water seepage through the roof coverings.
6. whether the inside of the roof is lined with felt, paper, lath and plaster or torching.
7. the roof junctions with any chimney stacks, especially axial stacks.
8. all flashings and fillets.

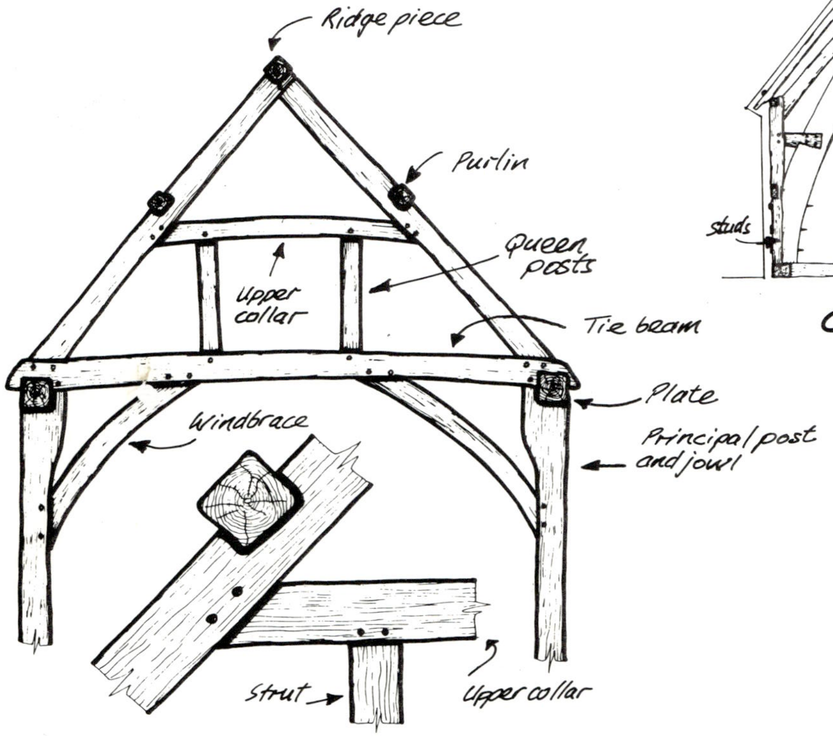

King Post Roof

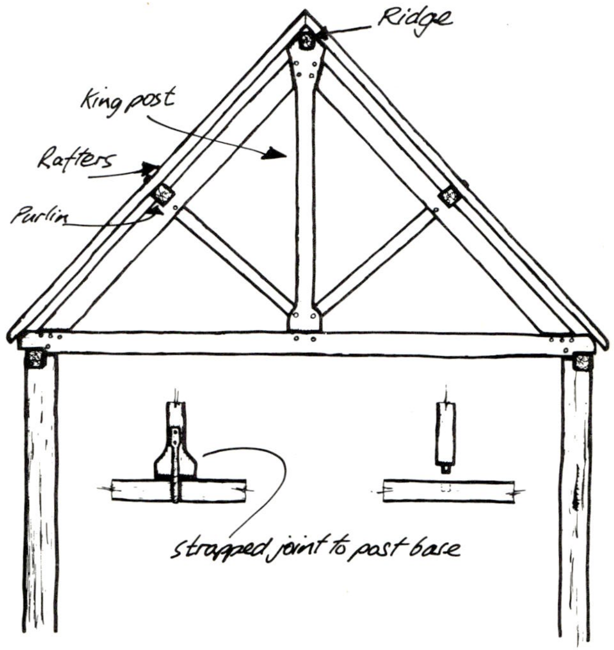

Ridge

King post

Rafters

Purlin

strapped joint to post base

Cruck Frame Roof

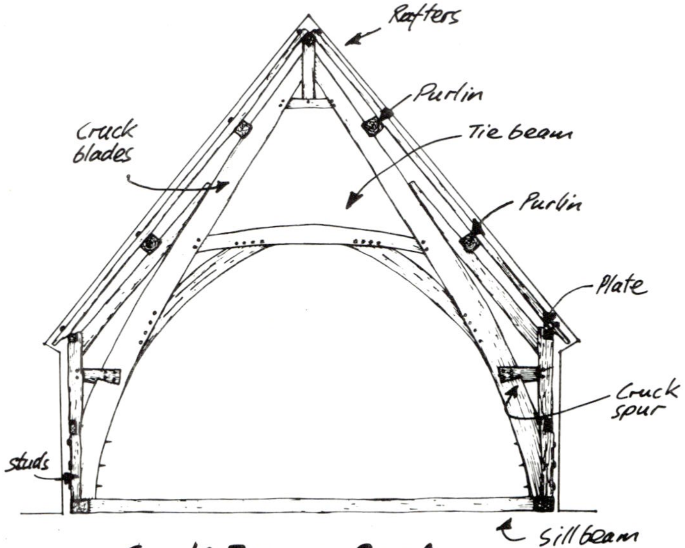

Rafters

Purlin

Tie beam

Purlin

Plate

Cruck spur

Sill beam

Cruck blades

Studs

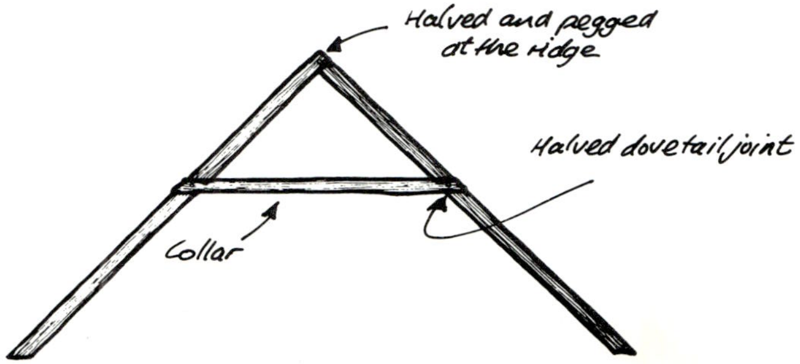

Single Collar Rafter Roof

Halved and pegged at the ridge

Halved dovetail joint

Collar

Queen Post Roof

Ridge piece

Purlin

Queen posts

Tie beam

Plate

Principal post and jowl

Upper collar

Windbrace

Strut

Upper collar

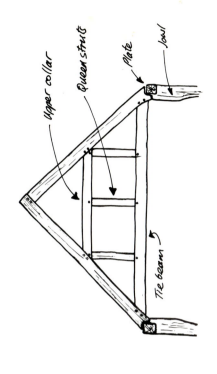

Upper collar
Purlins
Struts
Tie beam
Principal wall post with tapered jowl
Rafters
Queen posts or struts
Plate

Queen Post Roof

Upper collar
Queen struts
Plate
Jowl
Tie beam

Simple Queen Post with 3 struts

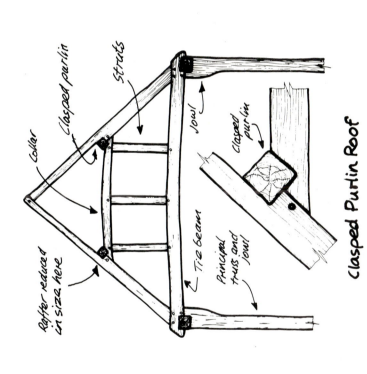

Collar
Clasped purlin
Struts
Jowl
Tie beam
Principal truss and jowl
Rafter reduced in size here
Clasped purlin

Clasped Purlin Roof

Look for

1. evidence of plate failure.
2. scarfing and joint decay.
3. major hogging in timbers.
4. evidence of differential roof thrust.
5. evidence of historic repairs in metal such as straps, ties, braces, etc. Check these for failure and rusting or acid attack.
6. signs of failure in any other metal fixings not part of the original structure.

Roof coverings

Are the roof coverings
1. plain tiles?
2. thatch?
3. slate and stone?
4. sheeted?

PLAIN TILES

Consider
1. the size of the tile.
2. whether it is machine made or hand made.
3. whether it is nib or peg.
4. which type of ridge it has.
5. which type of mortar was used.

Check
1. signs of tile slippage.
2. evidence of structural spread.
3. evidence of hogging and batten failure.
4. evidence of rafter sag and failure.
5. water ingress to chimney stacks.
6. the type of flashing/fillet.
7. the junction of eaves tiles with the wall plates.

Look for signs of
1. lamination.
2. broken/chipped tiles.
3. moss and lichen attack.
4. disturbance of flaunchings and undercloakings.
5. jack rafter movement.
6. sarking felt lining.

Half round clay ridges

Tile and a half

Tile battens

Common rafters

Bonnet tile

Peg tile

Nib tile

Plain Tile Roof Coverings

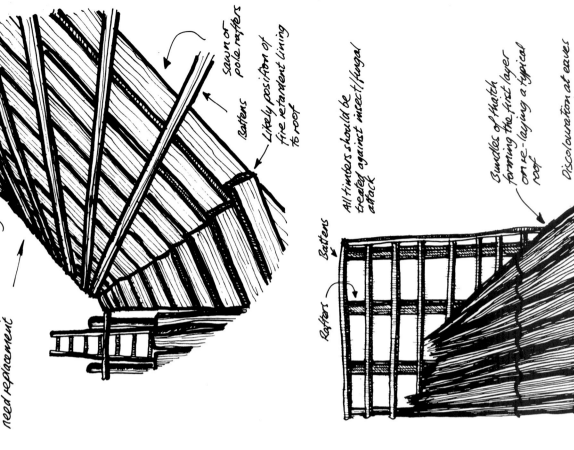

If the thatched roof is stripped down, look for evidence of insect attack and weak pole rafters which may need replacement

Sawn or pole rafters

Battens

Likely position of fire retardant lining to roof

All timbers should be treated against insect/fungal attack

Rafters

Battens

Bundles of thatch forming the first layer on re-laying a typical roof

Discolouration at eaves due to weathering

Typical Thatch Details

Roof coverings

*If there is **felt lining***

    does it

Are gutters provided?

    Do they

1. overhang the rainwater gutters?
2. project beyond the lower face of verge tiles?

1. align correctly to the eaves?
2. dispose of water from the hipped or flying hipped ends?
3. fall to proper intact downpipes?

*Are the **downpipes***

    connected to

1. open shoes?
2. proper soakaways?

THATCH

Is the thatch

1. Norfolk Reed (Water Reed)?
2. Combed Wheat Reed (Devon Reed)?
3. Longstraw?

Identify and discuss
lifespan and servicing

    Typically

1. Norfolk Reed's 60-year lifespan with 10 to 12 years between services (re-ridge).
2. Devon Reed's 40-year lifespan with 10 years between services (re-ridge).
3. Longstraw's 25-year lifespan with 8 to 10 years between services (re-ridge).

Roof coverings

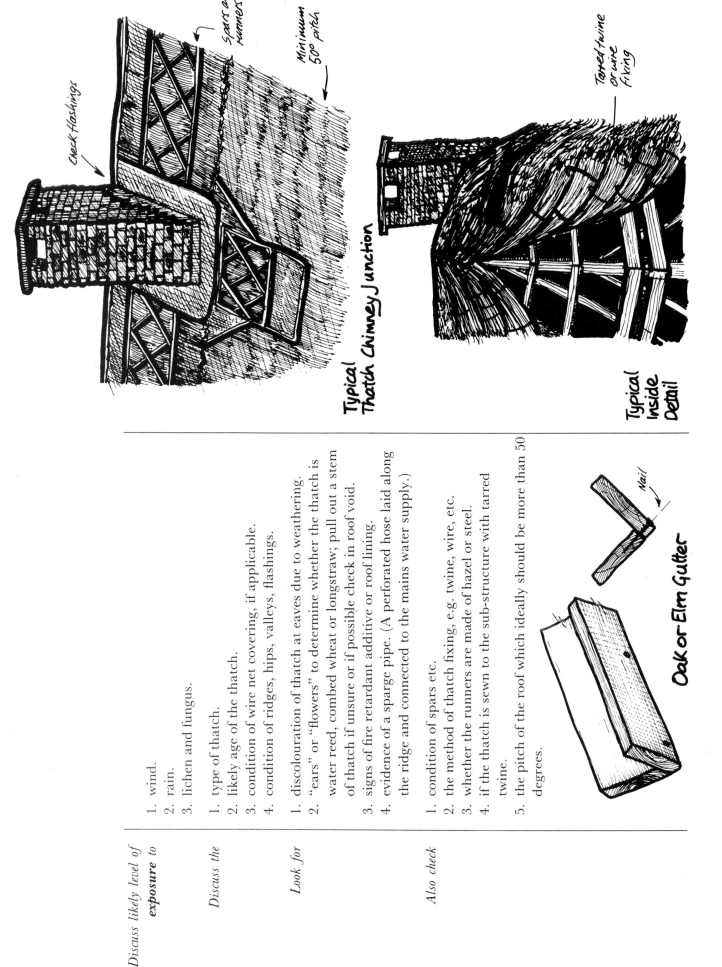

Check Flashings

Spars and runners

Minimum 50° pitch

Typical Thatch Chimney Junction

Tarred twine or wire Fixing

Typical Inside Detail

Oak or Elm Gutter

Nail

Discuss likely level of **exposure** *to*
1. wind.
2. rain.
3. lichen and fungus.

Discuss the
1. type of thatch.
2. likely age of the thatch.
3. condition of wire net covering, if applicable.
4. condition of ridges, hips, valleys, flashings.

Look for
1. discolouration of thatch at eaves due to weathering.
2. "ears" or "flowers" to determine whether the thatch is water reed, combed wheat or longstraw; pull out a stem of thatch if unsure or if possible check in roof void.
3. signs of fire retardant additive or roof lining.
4. evidence of a sparge pipe. (A perforated hose laid along the ridge and connected to the mains water supply.)

Also check
1. condition of spars etc.
2. the method of thatch fixing, e.g. twine, wire, etc.
3. whether the runners are made of hazel or steel.
4. if the thatch is sewn to the sub-structure with tarred twine.
5. the pitch of the roof which ideally should be more than 50 degrees.

Stone coverings are very heavy—look
for evidence of frame overloading

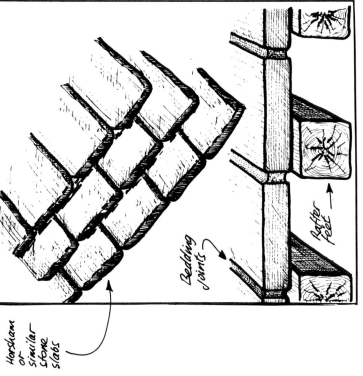

Horsham
or
similar
stone
slabs

Bedding
Joints

Rafter
Feet

Stone Roof Coverings

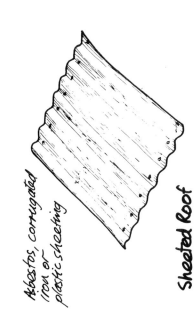

Asbestos, corrugated
iron or
plastic sheeting

Sheeted Roof

SLATE AND STONE

Look for evidence of

1. lateral spread due to overloading.
2. slate/stone slippage due to failure of fixings.
3. mortar pointing on stone roofs indicative of fixing failure.
4. nail sickness on slated roofs.
5. tingles.
6. decay to slate and stone surfaces from acid rain and gases such as carbon monoxide.

Check

1. all fillets and flashings.
2. all nail/timber fixings.
3. all structural abutments.

SHEETED ROOFS

Note

in some rural areas sheeted roofs can be found. These must always be regarded as temporary.

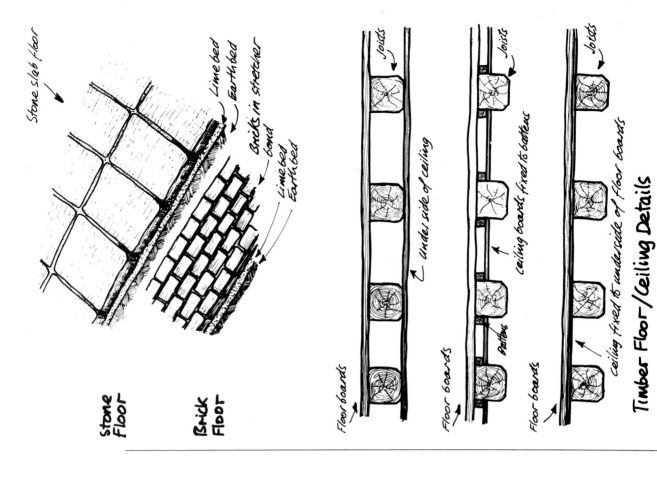

Stone slab floor

Lime bed
Earth bed
Bricks in stretcher bond
Lime bed
Earth bed

Stone Floor

Brick Floor

Joists

← underside of ceiling

Floor boards

Joists

Floor boards

Battens

ceiling boards fixed to battens

Floor boards

Joists

Ceiling fixed to underside of floor boards

Timber Floor/Ceiling Details

Floors

Check the construction

Is it

SOLID FLOORS

1. stone?
2. brick?
3. concrete?

Look for evidence of

1. dampness.
2. high ground levels.
3. slab settlement.
4. chasing.
5. different types of coverings.
6. condensation.
7. high alumina cement.
8. the use of a damp proof membrane (DPM).

TIMBER FLOORS

Timber ground floors

Check

1. the underside construction.
2. for the use of a slab.
3. for the use of a DPM.
4. for signs of joist decay and insect attack.
5. for signs of rising dampness.

By using a stressed cable and restraint, joists are drilled and supported in a crossbeam effect

Wall

Wall

Floorboards

Plastered ceiling structure

Cable Repair

Pipe

Pipe Chase Repair

Compression fitting to repair a pipe chase

Pipe

Fixing plate

Pipe

Typical Joist Repairs

Cut back decayed section and wedge

wedges

Rotted section

Wedge

Example of Timber Repair to a Rotted Joist

Suspended timber floors

Check for

1. bearing on external walls.
2. signs of lateral displacement of the walls and joist end decay.
3. evidence of metal or polymer repair.
4. timber repair.

Look at the

1. amount of floor distortion compared to wall/roof distortion.
2. gaps between floor boards and skirtings; comment if they are excessive.
3. general condition of floor boards.

Comment on

1. timber guarantees.
2. the need for re-fixing or replacement of boards.
3. general hogging and sympathetic movement.

Suggest that solicitors check all guarantees before exchange of contracts

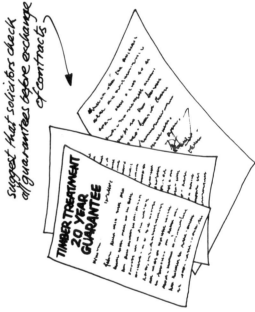

TIMBER TREATMENT
20 YEAR
GUARANTEE

Books for Further Reference

Addey, S.O., *The Evolution of the English House* (Allen & Unwin, 1898; revised 1933)

Alcock, N.W., *A Catalogue of Cruck Buildings* (Phillimore, 1973)

Barley, M.W., *The English Farmhouse and Cottage* (Routledge & Paul, 1961)

Braun, H., *The Storey of the English House* (Batsford, 1940); *Old English Houses* (Faber & Faber, 1962)

Briggs, M.S., *The English Farmhouse* (Batsford, 1953)

Brown, R.J., *The English Country Cottage* (Hale, 1979); *Timber Framed Buildings of England* (Hale)

Brunskill, R.W., *Illustrated Handbook of Vernacular Architecture* (Faber & Faber, 1971; new edition 1978); *Traditional Buildings of Britain* (Gollancz, 1981); *Timber Building in Britain* (Gollancz)

Building Research Establishment Report, *Recognizing wood rot and insect damage in buildings* (B.R.E. Department of the Environment)

Cave, Lyndon F., *The Smaller English House – Its History and Development* (Hale, 1981)

Clifton-Taylor, Alex, *The Pattern of English Building* (Batsford, 1962; new edition Faber & Faber, 1972)

Cook, Olive, and Smith, Edwin, *Old English Cottages and Farmhouses* (Thames and Hudson, 1954)

Cordingly, R.A., *British Historical Roof types and their members: A classification* (Ancient Monuments Society)

Crosseley, F.H., *Timber Building in England* (Batsford, 1951)

Cunnington, P., *How Old is Your House?* (Alpha Books, 1980); *Care for Old Houses* (Prism Alpha, 1984)

Fry, Eric C., *Buying a House?* (David & Charles, 1983)

Harris, Richard, *Discovering Timber-Framed Buildings* (Shire Publications, 1978)

Harrison, J.A.C., *Old Stone Buildings* (David & Charles, 1982)

Hewett, C.A., *English Historic Carpentry* (Phillimore, 1980)

Hollis, M., *Surveying Buildings. Third Edition* (RICS Books, 1990)

Innocent, C.F., *The Development of English Building Construction* (Cambridge University Press, 1916)

Lander, Hugh, *The House Restorer's Guide* (David & Charles, 1986)

Mason, R.T., *Framed Buildings of England* (Coach Publishing House)

Melville, I.A., and Gordon, I.A., *The Repair and Maintenance of Houses* (Estates Gazette, 1979)

Mercer, Eric, *English Vernacular Houses* (Royal Commission on Historical Monuments, 1975)

Peters, J.E.C., *Discovering Traditional Farm Buildings* (Shire Publications, 1981)

Powys, A.R., *The Repair of Ancient Buildings* (Dent & Sons, 1929; new edition SPAB, 1981)

Prizeman, J., *Your House – the Outside View* (Hutchinson, 1975)

Richardson, Stanley A., *Protecting Buildings* (David & Charles, 1978)

Royal Institution of Chartered Surveyors, *Structural Surveys of Residential Property – A Guidance Note. Second Edition* (RICS Books, 1985)

Ryder, Peter F., *Medieval Buildings of Yorkshire* (Moorland Publishing, 1982)

Salzman, L.F., *Building in England Down to 1540* (Oxford, 1952)

Society for the Protection of Ancient Buildings Technical Leaflets
  No. 1 *Outward Leaning Walls*
  No. 2 *Strengthening Timber Floors*
  No. 3 *Chimneys in Old Buildings*
  No. 4 *The Need for Old Buildings to "Breathe"*
  No. 6 *Introduction to the Treatment of Rising Damp*
  No. 8 *Treatment of Damp in Old Buildings*
  No. 10 *The Care and Repair of Thatched Roofs*
  No. 11 *Panel Infillings to Timber-Framed Buildings*
  No. 12 *The Repair of Timber Frames and Roofs*

Swindells, David, *Restoring Period Timber-Framed Houses* (David & Charles)

West, Robert, *Thatch – a Manual for Owners, Surveyors, Architects and Builders* (David & Charles)

West, T., *The Timber-framed House in England* (David & Charles, 1971)

Wodo, M.E., *The English Medieval House* (Phoenix, 1965)

Woodforde, John, *The Truth about Cottages* (Routledge & Kegan Paul, reprinted 1979)